The Culture Wars Globalists Vs Nationalists

Contents

Introduction

Culture is the determinant of what we value and thus trade, and therefore shapes what we refer to as "the economy". In a ground breaking analysis of the "Globalist" vs "Nationalist" debate in the U.S., Prof. Frank Karue Ngarua reveals the impact of these two visions on the culture of the population.

Prof. Ngarua defines culture as standardized actions that individuals take given various life scenarios. He categorizes cultures by usage and characteristics.

When categorized by usage, cultures can be divided into three main types; individual cultures (e.g. what time you choose to regularly go to sleep or wake-up), group adopted cultures (e.g. marriage, circumcision, symbolic rituals, traditions, etc.), and popular cultures (e.g. the current use of social media for dating, news, etc.).

Categorized by characteristics, there are two main types; "collectivist" and "individualistic". "Collectivist" cultures give priority to the needs of the collective society over the needs of the individual, while "individualistic" cultures do the opposite.

Prof. Ngarua's research shows that, based on historical evidence, "collectivist" cultures create populations with "dependency" and "conformist" cultural characteristics, who spend large amounts of their time on destructive distractions, in an attempt to escape their toxic reality.

Like all living organisms on earth, humans need the freedom and independence found freely available in the natural/natures environment. It is essential for a good quality life experience and for survival.

This lack of freedom and independence in "collectivist" societies, therefore, does have a significant negative impact on the quality of life of the individual and those around them. However, it does create revenues and profits for a few people, (i.e. "the elites"), increasing "income inequality" and creating a rigid class based system.

Today, the so called "Globalist elite" are the latest small group of individuals to create a system that promotes and or enforces "collectivist" cultures on large portions of a population, which has historically always ended in a collapse of the system, many a times violently.

With his new more precise definition of culture, Prof. Ngarua creates a paradigm shift in the understanding of the impact of "cultural" or "behavioural" economic visions, plans and strategies on the lives of today's populations worldwide. The reader should keep this definition in mind as they evaluate the cultures discussed in this book.

Whereas the "Globalist" vision emphasizes the empowerment and dominance of institutions in the economy, the "Nationalist" vision promotes the empowerment of the individual.

The "Globalists" envision a world where a group of "elites", via the institutions they control, make choices for the population in all spheres of their economic and social lives. This vision is currently being implemented using methods and tactics used in "Communist" and "dictatorship" led nations, via the institutions they control. The implementation aims to eliminate self-confidence in an individual's ability to provide for themselves independently, by sowing self-doubt in ones skills and abilities as well as their perception of self-worth.

The "globalist" elite are threatened by technological advances and easy access to information. Whereas, these two developments have given them certain advantages including less labour costs, which have increased their profits, the same development has given the population the ability to be less dependent on institutions and to ultimately have the ability to compete effectively in business with the institutions.

The anti-competitive strategies that the institutions have practiced traditionally are the reason for the "globalist" elite's wealth advantage. This disruption by technology that allows the population using small businesses to compete with the institutions and provide the same quality goods and services would radically re-distribute wealth away from the "globalist" elites who have used the monopoly power of the institutions to enrich themselves. This is why they are fighting so hard and have come up with an agenda that aims to stop this progression.

The "Nationalists" vision supports the individuals freedom to choose, and would like to see the freedoms enshrined in the U.S. constitution; such as the right of citizens to vote, bear arms, and free speech; preserved. They also support competitive markets over monopolies and are challenging the current domination of a few companies in many crucial industries.

The motives of the opinion leaders in the two camps is irrelevant, and has been used as a diversionary tactic, especially by the "globalists". The implementation of their visions is what counts, for it will shape the culture and thus economics of the U.S. population for years to come.

Ultimately, the responsibility lies with the population to see the "globalist" agenda in light of the impact it will have on their culture and life-style choices and make decisions appropriately. Individuals need to value being in control of their destiny, for this is what is at stake. The "globalists" want to create a new kind of slavery, where the population's choices are made by an elite "globalist" group. This book analysis the "globalists" agenda and gives the population a chance to choose wisely.

Chapter 1 – How we got here

Before the "industrial revolution" people grew their own crops, hunted for their own meat, gathered wood for their energy needs, built their shelters from materials they collected themselves from the land and provided their own security. Essentially, they provided for themselves independently by harnessing the freely available resources on earth.

During this period, tools that allowed people to provide for their needs were innovated and the knowledge on how to make these tools widely shared. For example, the know-how on innovations such as; fire, knives, cultivation hoes, ox plows, etc., was widely shared and not, for example, patented to prevent others from using the know-how to create their own tools from the natural resources available.

In the above environment, individuals were empowered to produce their basic needs for themselves, without having to trade in order to get them. Trading was therefore not a necessity, and available documented history shows that when trading begun, it was mostly based on goods and services that were not essential for the people's basic survival/needs. The now famous "Spice trade" occurred during a period when most people provided the basic necessities of life for themselves without trading (i.e. either their labour or goods), and most traders were doing it as a part-time venture, while still practicing subsistence farming within their homesteads.

The "industrial revolution" changed this arrangement drastically. Today most people's basic survival needs, i.e. food, shelter and security; are now being sourced from institutions. In order to source (i.e. buy) these products and services from the institutions, you must trade something of value with them and traditionally this has been your labour/skills. People therefore changed their skill set from one of self-sustenance (hunting, gathering, cultivating the land, etc.), to the kind of skills the institutions were looking for, in order to trade with these institutions. They slowly became entirely dependent on these institutions for their basic needs as a result.

During the same period, innovations and their relevant know-how or knowledge became an increasingly valuable commodity to be traded. As a result know-how on tools, manufacturing processes, etc., became trade secrets to be protected. This ensured that many tools and products could only be sourced via the institutions by trading your labour with them first for money. Patents, copyrights and trade-marks protected this knowledge for exclusive use, with the initial intention of providing motivation to inventors, but instead it soon became a tool wielded by institutions to keep out competition and ensure dependability of the population on the institutions. The population's ability to use natural resources to produce for themselves newly discovered goods and services became difficult if not impossible in most cases.

Chapter 2 – The "Nanny Corporation"

In order for the new industrial institutions to get the labour they needed, they had to increase the attractiveness of the trade economy. They did this by providing employment stability, free or subsidized housing, paid days off, paid health care, etc. Essentially the "Nanny Corporation" was born.

At the same time, since the people were trading their labour for money, new auxiliary businesses such as grocery stores, restaurants and construction companies emerged to provide the needs of the employees, since they no longer had the time to provide these needs for themselves. Many of these new businesses also grew big and employed many people, increasing the dependency of the population on institutions.

This created the current large population that is totally dependent on institutions for their basic survival needs. This new environment required a different culture; one which was compatible with the new realities for the vast majority of the population. The new reality was that of a population, fully dependent on institutions to survive, and which therefore had to conform to the expectations of their new "overlords".

Chapter 3 – "Globalization"

As the industries grew in the "developed" countries, capital and labour shortages began to emerge due to border and immigration controls. These shortages were leading to higher costs of doing business and thus less profits for the institutions owned and controlled by the "globalists".

Free movement of goods and services, commonly referred to as "free trade" was the first agenda advocated by the "globalists". Nations were encouraged to specialize in the production of certain goods and services and trade with other countries to acquire the goods that they did not have an advantage in producing. This is known in economic jargon as "competitive advantage". The "World Trade Organization" was tasked with implementing this agenda and successfully made the new reality we currently live with today. Many of the tariff barriers previously erected to protect local industries within national borders in order to encourage self-sufficiency for national security reasons, were removed to allow the free flow of goods. Countries became reliant on certain goods and services coming from outside their borders, losing local expertise and know-how in various industries that no longer existed within their borders. As a result, these countries lost the capabilities to internally produce those goods and services. Many manufacturing companies started relocating their plants to developing countries for cheaper labour, despite the length and logistics of moving the finished goods back to the developed countries for consumption. Many manufacturing jobs were lost in the developed countries, but many were retrained with skills needed in the growing "service industry". What was not discussed is the fact that the "service industry" is a derivative (or support industry) of the manufacturing industry, which in many cases is usually geographically located near the industry it supports.

Next, came the agenda to allow the free movement of money between countries for investment purposes. Many countries had strict policies ensuring their citizens had priority, when it came to investing in businesses domiciled in their countries. There were also various foreign exchange restrictions around the world that protected local currencies from fluctuations that could negatively impact the local economies. The "globalists" slowly eroded these policies, promoting what is now referred to as "foreign direct investment", and emphasizing its success in growing various national economies around the world. Local currency became a tradable commodity on Foreign Exchange markets, allowing it to fluctuate in value Vis-a Vis other currencies, and exposing the currencies to potential manipulation.

Then came the "globalists" agenda, which caused considerable friction with large portions of the populations in the developed world, the free movement of people without border restrictions. This was the first warning in the developed world that the "globalists" agenda could be harmful to their way of life, especially economically.

One of the major reasons why internationally recognized national borders and immigration laws were put in place was to address the problems caused by the domination and colonization of large parts of the world by peoples from the more militarily and technologically advanced countries or communities. The domination and colonization had devastating effects on the defeated communities, including the loss of choice, natural resources, as well as slavery.

These border controls, however, were now a hindrance to the "globalists" quest to deal with the shortages of labour in "developed" countries by moving populations from the "developing" world to the "developed" world and thus decreasing wages and raising their profits. The "globalists" therefore became advocates of what are now commonly referred to as "open border" policies.

To gain the support of populations in the "developing" world, they framed the "open borders" campaign as a one which supported the economic well-being of these populations, and inserted a racial angle to the debate as an additional incentive.

The "globalists" promoted the dependency of both the populations and the governments of these "developing" nations on money remittances from their populations, now relocated to the "developed" world. This destroyed their ability to create internal value and trade. The populations in these developing countries became increasingly dependent on their governments, foreign aid, foreign direct investments, foreign remittances from their relatives, and numerous foreign charities. They slowly adopted a dependency culture.

In her New York Times bestseller book, Adios America, Ann Coulter describes how Mexican Billionaire Carlos Slim's businesses have benefited from the remittances of Mexican's, who work in the U.S. She goes on to say that "Immigration isn't about rescuing the 2.4 billion people of the world living on less than two dollars a day. It's about enriching the already rich, who like to laugh at blue-collar people being ground down by cheap labour". She goes on to state that "The elites just can't grasp that most American's raise their own kids, make their own beds, cook their own food and drive their own cars." According to Wikepedia.com, Carlos Slim is the richest man on earth and approximately 40% of the stocks on Mexico's stock exchange belong to companies he owns.

For many years, the elite in the U.S. envied their counterparts in developing countries that cheaply employed maids, cooks, gardeners and drivers. It was very expensive to do this in the U.S. until the "open borders" policy brought in waves of cheap labour starting in the 1990's. This is one of the incentives that has encouraged many "elites" and "want to be elites" in the U.S. to support open border policies. However, the negative repercussions of the policy are starting to change the minds of many.

Chapter 4 – Why the "globalists" need us

Without labour and consumption from the population, the institutions cease to exist. Technological advances have greatly reduced their need for human labour, however they still need the population to consume their goods and services. This simple fact is the reason the institutions fight so hard to eliminate any progress back to self-sustainability.

The institutions are losing this fight because of a drastic change in the economic environment. Specifically, the new economic environment has short new product/new skill/new innovation life cycles and fast changing consumer needs which results in "brutal" competition, smaller profit margins, and falling wages. In this new economic environment, businesses that enable the population to empower themselves and move back towards self-sustainability are flourishing. Companies such as Microsoft, Amazon and Google started with this goal of empowering individuals, with Microsoft providing productivity software (MS word, excel, etc.) that allowed individuals to use productivity tools previously only found within the institutions, Amazon allowing individuals to set up their own online stores on their platform and Google revolutionizing the way information was accessed. These companies grew rapidly by adopting this strategy, and now are among the biggest companies in the world. Today, however, they have slowly started changing their strategy to one of using their dominant positons to eliminate competition and disempower the customers.

To counter this reality, the institutions have strategically positioned themselves to look like enablers, while in-fact providing products and services that still keep you dependent.

Here is an example. With the computerization of many of the tasks that were previously done manually using human labour, the institutions are fighting hard to ensure they dominate this sector and provide resistance to the potential devolution of knowledge and capabilities. A good example is the computer software industry which increasingly only allows users to access its products on a remote server (the so called "cloud computing"), ensuring that you can never own the software but are simply leasing it. The old software purchasing business model where the product actually resided on the purchaser's computer and could potentially be used by the owner forever is slowly becoming a thing of the past. Even more empowering, would be a population with the ability to create their own productivity software on their computers for themselves and potential sale to others. This would create many small and medium enterprises, enhance competition and motivate innovation in the software industry. Instead, a hand-full of software companies dominate and only allow consumers remote access on a lease basis.

The new economic environment is a perfect "breeding ground" for a devolved economy that will bring competition and innovation from small and medium enterprises, with no dominating institutions. This would also eliminate corruption (crony capitalism, etc.), which has survived in the anti-competitive environment controlled by a few dominant institutions.

Those that benefit from the current status quo are fighting hard to stop this natural economic progression. The strategy involves mergers and acquisitions to eliminate competition, large legal departments to protect their intellectual property and protect themselves from charges of theft of the same and open borders to keep labour costs down in order to maintain inflated wages and privileges for a small group of elite.

Chapter 5 – The "globalists" plan

The "globalist elite" plans to create a "robotic" human population whose thoughts and actions they control completely. Their success so far is stunning and disturbing. Large portions of the population use "packaged thoughts" (more on this later in the book) and clichés in every day debates on issues. Their ability to come up with original thoughts has been greatly diminished. Many use robotic actions in their every-day life, similar to employees on an assembly line, where they perform various tasks in a set way without much thought and when they encounter a new task have difficulty adapting. Even when talking, many today talk fast, and rarely analyse or evaluate what they are saying, which mostly comes from "packaged thoughts and ideas". The saying "think before you speak" is slowly becoming a relic of the past! Even before one finishes asking a question, many people today will begin their reply, many a times a reply that is not relevant to the question asked! In another example that shows the absurd nature of today's popular culture, if you pause to think before answering a question, the person asking the question will quickly say "never mind", and walk away! The "robotic" life where people go through life without paying much attention to their surroundings, not asking questions, and thinking/analysing little if anything about their lives and circumstances is slowly becoming a reality.

A small group of financially wealthy individuals, now commonly referred to as the "globalist elite" is trying to mould the world into an image that benefits their financial wealth. At the heart of their problem is the fact that technology is projected to eradicate the vast majority of jobs, which provided the income for their customers. They like the increased profits accrued from the savings of using technology rather than human labour, however, their businesses need customers, and in the current economic structure, without jobs, there can be no customers! This dilemma has got them strategizing on various solutions, now commonly referred to as the "globalisation agenda".

Since the "globalist elite" no longer need labour as much, but still need consumers, certain attributes of the current human are problematic. Attributes such as acquisition of technical skills and analytical thinking were useful to the "globalist elite" for the labour of their businesses, but have now turned into a problem for the elite. The rapidly increasing job-less population, is using their technical skill and analytical thinking to compete effectively against their businesses, and utilizing the technological advances to their advantage! This scenario is increasingly eating away at their wealth, which will greatly affect the kind of life style they are accustomed to.

You-tube videos are a good example of how individual empowerment is impacting the "globalists". Now commonly referred to as You-tube University, many are turning to "how to" videos to learn how to provide certain goods and services for themselves outside of the institutions. Some of the most popular videos are videos on how to start your own business online, how to cook, how to repair various household items, etc. 3D printers are also threatening the manufacturing dominance of the institutions, including printers that can produce guns.

In an amazing twist of fate, the "globalist elites" are turning to tactics and methods used in "communist" and "dictatorship" led nations, to eliminate their perceived threat from a technically skilled and thinking population.

After years of being militarily and financially protected from nationalization and wealth confiscation policies adopted by the "communist" and "socialist" leaning nations with taxpayer funds from "capitalist" leaning countries, they are now using their dominant ownership/control of media, communication channels and educational institutions in these countries that protected them, to eliminate independent thinking, free speech and the culture that values work over laziness.

Theodore Dalrymple observed the following about the current state of affairs, "In my study of communist societies, I came to the conclusion that the purpose of communist propaganda was not to persuade or convince, not to inform, but to humiliate; and therefore, the less it corresponded to reality the better. When people are forced to remain silent when they are being told the most obvious lies, or even worse when they are forced to repeat the lies themselves, they lose once and for all their sense of probity. To assent to obvious lies is...in some small way to become evil oneself. One's standing to resist anything is thus eroded, and even destroyed. A society of emasculated liars is easy to control. I think if you examine political correctness, it has the same effect and is intended to."

Nationalization and confiscation of wealth has now become very rare even in the communist countries such as China, and therefore the "globalist elite" no longer have a fear of this happening much anymore. Today they are therefore working and collaborating with "communist" and "socialist" countries to push their "globalist" agenda in the countries that formerly protected them from these very nations!

There has been talk of a "universal income" as a bait to get the population to buy into their vision and provide support. Other baits include "open borders" which appeals to many in mismanaged and corrupt countries where opportunities are scarce, the creation of a "want to be" elite class thriving on an illusion that they and/or their children can join the elites and live a life free of manual and/or physically challenging work, and promises of "rights to basic necessities". The campaign has similarities to the way companies enter a market via dumping of goods and services, with artificially low priced products to undercut their

Predatory lending practices have ensured that most citizens will die before paying off the mortgages/loans on the houses or cars they "own". Inflated college fees driven by generous student loans from the federal government, that have allowed employees in many colleges to earn salaries abnormally high in an economy where 75% of newly created jobs pay between $10-15 an hour, have ensured that most graduates will spend most if not all of their lives paying off their student loans. A culture of teamwork (similar in design to communal farming in communist countries) introduced into the U.S. corporate environment in the 1990's has ensured individual achievements are no longer valued, and an environment where tactics to get others to do more than you, are the most prized possession. A vast majority of the U.S. population today lives beyond its income, with debts exceeding their assets, making them vulnerable to the institutions controlled by the "globalists".

competitive edge. The human body, for example, is very fragile and weak when compared to other carnivorous animals, but that is compensated for by the mind which has enabled humans to build tools that mitigate this weakness.

The danger today is that if the "globalist elite" have their way and structure the world in their preferred image, the vast majority of humans will no longer possess this competitive advantage and neither will they possess the attributes that mitigate this weakness. They will be fully dependent of the "globalist elite" ensuring that they can leverage this vulnerability for their profits.

Evidence of this can already be seen currently in the world, where the "business of poverty" is the fastest growing and dominant economic sector in the world. It is only challenged by the "business of distractions", which is helping eliminate the ability of humans to analytically and independently think for themselves.

Already, many are afraid to give opinions contrary to those encouraged by the "globalist elite" due to repercussions built into the subjective system of granting access to the opportunities controlled by the "globalist elite" who currently dominate the world economy. Others avoid thinking by being distracted (see "culture of distractions"), and or adopting robotic speech (talking without thinking much and using packaged speech they have learnt verbatim). Other people have adopted robotic actions (much like employees in the fast food restaurants that are trained to do their jobs using robotic like actions with little thinking), and transferred them to their daily life's actions including driving of vehicles!.

If the "elite globalists" succeed in their plan, it is unlikely to be sustainable in the "long run", meaning that when it collapses, there will be a human population without its thinking capabilities competitive edge, and without the competitive advantages that help other animals survive without much thinking power. This may result in the human population evolving after a long period of time back-wards to reacquiring competitive advantages needed by animals without advanced thinking capabilities, reacquiring their thinking competitive edge if it has not been too seriously damaged, or becoming an extinct species.

In his book, 1177 BC, the year civilization collapsed, Eric Cline discusses the similarity between what is happening today, and the period he discusses which ushered what is referred to today in history as the "dark ages". He cites an article published in 1998 by Susan Sherratt, which points out that the collapse ended the "old centralized politico-economic systems" of that time (referred to in history as the "bronze age") and ushered in a decentralized economic system (referred to in history as the "iron age"), with smaller city states and individual entrepreneurs who were in business for themselves.

Today's economic environment is ripe for economic and political decentralization, but this a threat to the "globalist elite" whose power is solely based on a population dependent on the institutions they control and benefit from. This explains their attempt to mould a different image that benefits them financially allowing them to continue to maintain their current life-style. This essentially, is what is at the heart of this battle. Centralized economic and political systems are the levers that allow the "globalist elites" to maintain their wealth and power.

During the period that was commonly referred to as the "cold war", the two most powerful countries militarily pursued different economic models which had been labelled "capitalism" and "communism". Other countries around the world aligned themselves closer to one or the other, but practiced various variations of these economic models. The main difference between the two models was in regards to who got to decide which goods and services would be produced. Whereas, in the "communist" economies, a small elite that controlled the monopolies made the decisions, the "capitalist" model emphasized competition among many providers allowing the population to make that decision. However, after the cold war was over, large institutions in the "capitalist" countries slowly became monopolies, and removed this power from the population, creating a "communist" economic model in the "capitalist" countries. Today, in many crucial industries such as communication, health, agri-business, media, etc., there are dominant institutions that control two thirds or more of these markets, ensuring they are the decision makers, and not the population. These institutions, which during the cold-war had enjoyed the protection (militarily and financially) of the "capitalist" countries from being nationalized in many countries, have now turned against the very population that funded these protections and imposed a "communist" economic model on them.

They have also ensured, just like in the former "communist" countries, that they provide directions to the population on how to think and act by punishing those who don't conform via the loss of economic opportunities. The culture that the population is now pressurized to conform to, is the main topic of this book. I discuss how the institutions are promoting certain cultures, how they enforce conformance, and how it benefits them financially.

Chapter 6 – Why the U.S. is central to the plan

The U.S. is the world leader militarily, economically and culturally. This makes the U.S. the most important target in the "globalization agenda". Without changing the U.S., their mission will surely fail.

This has proved to be a very difficult and challenging task for the "globalists". The U.S. is a country where certain individual freedoms were inserted into the constitution during its independence, by a colonist/settler population that was familiar with the aspirations of an oppressed European population expressed via writings during an era now commonly referred to as the "Enlightenment" period in Europe. Examples of these freedoms include the right to bear arms.

This right is enshrined in the Second Amendment of the U.S. constitution and according to Wikipedia; "The Second

Amendment was based partially on the right to keep and bear arms in English common law and was influenced by the English Bill of Rights of 1689. Sir William Blackstone described this right as an auxiliary right, supporting the natural rights of self-defence and resistance to oppression, and the civic duty to act in concert in defence of the state."

As a result of its constitution, and the willingness of large portions of its citizens to constantly defend individual freedoms, the U.S. is unique. This is also the reason that many around the world consider the U.S. an opinion leader in many aspects of human life and endeavours.

It may also be the clearest sign that diversification of economic, political and social power via allowing individuals the freedom to choose is the best formula for improving the living conditions in a community or nation. In the U.S. the role of institutions in the citizen's lives has always been a subject of debate, with many supporting a limited role. The motto of the U.S. state of New Hampshire says it all; "Live free or die"!

Chapter 7 – The economics of poverty

The "globalist" elite want a class based society that is structured like a pyramid scheme, much like the one that existed when "landlords" and "royal families" owned and controlled all means of production, thus creating a poor underclass that was dependent on the elites and thus disempowered and vulnerable. The now commonly popular phase that many use, that they wish they could be in a position where "their money works for them" is very much in line with this pyramid scheme philosophy. By getting to the population to accept that it is okay to enrich oneself via a pyramid scheme strategy, they will see nothing wrong with what the "globalist elites" are doing, and instead will be inspired to try and accomplish this strategy themselves. This philosophy has been spread via the media and education systems which are now mostly controlled by the "globalist" elite and their sympathisers. The increased propaganda showing royal families and elite bourgeoisies as a class of citizens that does many good deeds for the impoverished is part of this strategy. They are eager to feed the poor with fish, while ensuring they can never learn how to fish for themselves.

The "globalist elite" have highlighted the value of charity and giving, while promoting strategies that disempower the recipients in order to keep them poor and dependent on charity. Organizations involved in charity work are one of the fasted growing economic sectors.

According to statista.com, donations to various charity institutions in 2016 amounted to $390 billion. According to Forbes magazine, the ten biggest charities had revenue of over $20 billion in 2016, with United Way, Feeding America and Salvation Army having revenues of about $3 billion each, much of which is spent on buying processed unhealthy foods from large institutions for the vulnerable populations who no longer have a choice in regards to their sources of basic necessities and other commodities.

According to techreport.ngo, there are over 10 million charity related organizations worldwide, whose combined revenues would make them the fifth largest economy in the world. The website also reports that there are 1.4 million charity related organizations in the U.S. that employ 11 million Americans. In India, there are 3.3 million charity related organizations, approximately one charity organization for every 400 people.

There has also been an increase in what is known as "slum" tourism. This is where the financially wealthy class tours slum areas, especially in the developing world, taking pictures of the poverty conditions. They use them, at times, as lessons to their children on how "privileged" they are (and most likely also as a warning to conform to the narrative of the "globalist elites"), as well as for raising funds for the charities they own and control.

Over the last three decades, a new "economic class" has emerged that benefits from others being dis-empowered. They support the efforts of the institutions to use various methods to thwart any attempts by the population to reduce their dependency on the institutions which they benefit from financially either via employment or ownership. Members of this class are expected to adopt and promote narratives, ideas and cultures that support the "globalization" agenda and continued dependence on institutions by the population.

Chapter 8 – The economics of distractions

Providing distractions to a disillusioned population has become big business. These businesses are some of the fastest growing in today's economy. A growing number of people around the world are willingly seeking various distractions to avoid the realities of their lives. The evidence is in the amount of time they spend on the various distractions, which is an indication of an addiction. The most common distractions include; religion, rituals, ceremonies, sex, gossip, drugs, sports, entertainment and alcohol. Many people engage in these distraction activities even when at work, causing many employers to put in place procedures and rules that attempt to stop these behaviors. On the road, distracted driving is widespread, endangering the safety of others. And when walking, it is now common to see people bumping into others and even tripping and falling because they are not paying attention to what is in-front on them, but instead are staring at their smart phone screens! Addiction is widely considered to be an indication of a person's attempts to avoid dealing with a perceived problem.

Many businesses have capitalized on this phenomenon, providing the distractions sought for a fee. In his best-selling book, "Empire of Illusion", Chris Hedges discusses how sports and entertainment are used to provide an alternative reality for many, helping them conquer (in their imaginations) their perceived obstacles via the actors and sportsmen. In reality, however, their perceived obstacles and problems remain, causing them to increasingly spend more time in their alternative reality. This works out well for the businesses providing these distractions, as their revenues grow.

Here are some examples of businesses that are benefiting from the growing culture of distractions:

According to an article on Forbes.com titled, "Sports industry to reach $73 billion by 2019", the sports market which was worth $60.5 billion in 2014, will be worth $73.5 billion by 2019. This is mostly because of "media rights deals", which means the number of people watching sports via various media is growing rapidly. Licensing of sports clothes via royalties by the owners of the sports team was a $698 million from the sale of clothes worth $12.8 billion in 2014 according to another article on Forbes.com titled "Sports licensing soars to $698 million in royalty revenue".

According to statisticbrain.com, the 11,000 adult films are released every year in the U.S. and the annual industry revenues now stand at $13.3 billion. 55% of guests in hotels in the U.S. will order adult films, while 87% of college students in the U.S. have had phone or webcam sex. Also, the number one internet search term is sex.

According to hootsuite.com, social media advertising budgets worldwide have doubled from $16 billion in 2014 to $31 billion in 2016. The author projects a

26.3% global increase in 2017. Social media content can be categorized as mostly entertainment, sports and gossip within personal networks.

Marketresearchstore.com in its 2016 publication titled, "Global depression drug market set for rapid growth, to reach around 16.80 USD billion by 2020" notes the following: "Zion Research has published a new report. According to the report, the global depression drug market was valued at USD 14.51 billion in 2014 and is expected to generate revenue of USD 16.8 billion by end of 2020, growing at a CAGR of 2.50% between 2015 and 2020".

According to an article on economist.com titled "Celebrities endorsement earnings on social media" according to captiv8, an analytics platform that connects brands to social media, "someone with 3m-7m followers can charge, on average, $187,500 for a post on YouTube, $93,750 for a post on Facebook and $75,000 for a post on Instagram or Snapchat".

There is also a direct relationship between the rise of ceremonies and rituals in a community and the progressive decline of the majority of the people's ability to provide for themselves. During the transition period, when the decline begins, ceremonies and rituals increase rapidly as the "globalist elite" beneficiaries of the system try to distract the majority from analyzing the cause of their declining fortunes.

Chapter 9 – The economics of complexity

The new economic environment has short new product/new skill/new innovation life cycles and fast changing consumer needs which results in "brutal" competition, smaller profit margins, and falling wages. In this new economic environment, the relevance or real value of the goods and services traded becomes a focal point of many. However, many institutions are trying their best to avoid this reality by using anti-competitive practices and attempting to suppress the dissemination of knowledge and information that would erode the monopolistic advantages they have relied on for so long to survive. The devolved economy that is emerging is a threat to the wealth creation strategies of many institutions that have relied on monopoly advantages. This devolved economy, where many small businesses compete on a level playing field aided by emerging technologies and a free flow of information threatens to shift and re-distribute wealth from the "globalist" elite, and put it in the hands of small business owners in a more evenly balanced distribution. You can find many industries whose existence is heavily dependent on complicating things that could be made simple, in which case the industry would either shrink drastically, have greatly reduced profits/wages or cease to exist. These industries continue to have large revenues and their services continue to be critical. I will look at four dominant economic sectors to analyze this phenomenon; Health, Education, Information Technology and Justice.

Health

The content in all fields of education continues to grow rapidly, and in specific areas, namely the so called STEM areas (Science, Technology, Engineering and Math's) more complex. The complexity of the so called STEM courses discourages many from pursuing them. They have thus created a shortage in these fields, which raises the wages and profits in these sectors. Is this intentional? There are many who believe that if an educator knows their subject matter well, they should be able to explain it in a short and simple manner to anybody with a high school education. So what is wrong with the STEM courses? The inability of the educators in the STEM courses to present their material in a short and simple way could only mean that they either do not understand the material well enough to explain it simply, or they are creating an intentional shortage of personnel to keep wages and profits up in their professions. This is a problem in an economy that is headed in the opposite direction for most of the population, and puts their services out of the affordable price range. In-fact, in the U.S. today, most of the revenue in the expensive health sector comes from government funded programs like Medicaid, Medicare and government subsidized health insurance programs. Were it not for the government programs, this sector would collapse due to a lack of patients that can afford their services. These STEM programs need not be complicated, and should enable the mass spreading of their knowledge to decrease prices and make the services affordable based on the incomes of the vast majority of their potential customers. If this does not happen, the culture of complexity which is currently enriching them, will lead to their downfall, when the government can no longer afford to subsidize their customers. The U.S. health care sector revenues are over 1 trillion U.S. dollars, according to CSImarket.com.

Education

The content of education has grown rapidly, however, its utility is now frequently questioned. Is it possible that the growth of content is simply a way to charge students more money in an education system that sells it services in packages (i.e. Associates Degree or Degree) that require a set number of courses and content? How useful is this information to the student after graduation, and could some of it be acquired later through self-study?

In the U.S. most students over the last couple of decades have funded their education using government loans. With the changing economic environment, the default rate on these student loans is up according to an article in the Washington Post dated 28th September 2017 titled 'The number of people defaulting on federal student loans is climbing". Many college graduates in the U.S. no longer believe they will be able to repay their student loans, given their earnings potential in the new economic environment.

Is it time for colleges to change the amount of content, length of the programs and keep it short and simple? Alternatives are already being provided, including free online universities, self-learning books and You-tube videos some of which are greatly shortening and simplifying topics that take a whole semester to teach in colleges.

The education sector is made up of public and private schools. In the U.S., According to usgovernmentspending.com, public (tax and lottery) funded schools in the U.S. make up the majority with total spending in 2014 equaling $915 billion. The largest public school spending occurred in California $112 billion (12%), Texas $77 billion (8%) and New York $74 billion (8%). Private schools in 2014 had revenues of $54 billion according to statista.com.

Information Technology

In the U.S. the Information technology (IT) sector had 346 billion Euros in revenues while the European Union had 314 billion Euros in IT sector revenues. Many employers say there is a shortage of IT professionals. Is the content of the training presented in a complicated difficult to learn methodology? If so why? The wages and profits in the IT sector are also known to be very high compared to other sectors. The U.S. communication sector, a critical sector dominated by a few companies, relies heavily on IT products and services. When compared to other communication sectors around the world, it charges its customers very high fees for services. It is valued at $ 1.4 Trillion by statista.com.

Justice

Have you ever tried reading a law passed by the U.S. congress? These laws are long and difficult to understand! Is this intentional? Law schools in the U.S. are some of the most expensive, and consequently acquiring the service of a lawyer for your defense is equally expensive. However, trying to represent yourself in a justice system full of complex rules, laws and procedures, is very difficult if not impossible! Essentially, the culture of creating complexity has captured another industry causing inflated costs, for a population with declining wages and profits! The largest revenues are generated in the corporate legal departments ($160 billion according to statista.com). According to Statista.com, the Legal Service industry in the U.S. is expected to generate a total of $ 288 billion in 2018.

The culture of creating complexity is costing everybody, by inflating prices, in an economic environment of falling wages and profits. Those in these sectors that are currently benefiting from higher wages and profits may not view this as a problem, however, when the institutions (government, corporate, etc.), which derive their revenues from the general population can no longer afford the inflated prices, the financial collapse in these industries will be massive. Another bubble waiting to burst!

Chapter 10 – "Globalists" have already changed culture

To achieve dependency and vulnerability in the U.S. population, the "globalist elite" have engineered a culture change promoted and enforced via the institutions they control. The culture promotes lifestyle and behavioural choices that generate revenues for the institutions and increases the population's dependency on the institutions. During my research on this issue, I produced a series of short publications on my website, habarishujaa.com, known as the Kikulacho Cultures Series™. The series discusses popular cultures, promoted by the "globalists", via the institutions they control and/or own, that are having a negative impact on your financial and personal well-being. The word "Kikulacho" comes from the Kiswahili (Swahili) language, and literally translated into English means "what is eating you". However, in practice it is used in the Kiswahili language to mean "what is ailing you". A popular Kiswahili saying "Kikulacho Ki nguoni mwako" is the inspiration behind the name of the short publications series. The practical English translation of this saying is "what ails you is within you"! Essentially, by changing the population's culture, they have made it the source of its own problems!

In this chapter, I have included the short publications to clearly show how the "globalists" have already changed large portions of the population's culture, and how they are benefiting financially from this change. By engineering this change via a revenue and profit incentive driven process, they have used the free markets to achieve their desired agenda.

THE CULTURE OF "INSTITUTION DEPENDENCY" AND YOUR FINANCES

In today's world, most people are dependent on Institutions for their daily basic needs. Many people today look for employment/contract or business opportunities with institutions in order to provide for their basic needs. These institutions can be categorized into three broad groups, Government/Public institutions, Private/Business institutions, and non-profit institutions. The "industrial revolution" which brought manufacturing and other industries into our lives set in motion this increasing dependency. Before industrialization, people's way of life revolved around hunting and gathering of wild plants and animals and later clearing and cultivating land and domesticating animals. With the setting up of manufacturing industries, institutions became more prominent in peoples way of life, which has brought us to this point where nearly all our needs are met via institutions today.

Whereas before the "industrial revolution" people grew their own crops, hunted for their own meat, gathered wood for their energy needs, built their shelters from materials they collected themselves and provided their own security, today all these needs are now being sourced from institutions by an increasingly large portion of the population. In order to source/buy these products and services from the institutions, you must trade something of value with them and traditionally this has been your labor/skills. People therefore changed their skill set from one of hunting, gathering, cultivating the land, etc., to the kind of skills the institutions were looking for in order to trade with these institutions. Technology has however now disrupted this relationship that has worked for nearly two centuries resulting in the institutions needing increasingly less labor/skills from people. As a result we have a people with obsolete skills that the institutions no longer need and institutions who have a problem selling their products and services to a people with nothing of value to trade with them in return! Even worse, able bodied people; with no skills and/or natural resources that would allow them to provide for their basic needs independent of the institutions, have been reduced to non-productive members of society.

Below I have the revenues currently generated by the industries that today provide our basic needs:

According to the website calcbench.com, 15 companies control 80% of the food and beverage industry in the U.S., whose annual revenue in 2013 was $626 billion. Archer Daniels Midland Company controls 15% of the market, PepsiCo 11%, Bunge Limited 10%, Heinz 10%, Coca cola 8% and Tyson foods 6%.

According to the website statista.com, in 2014, the U.S. frozen ready dinner's market generated $15 billion in revenues.

According to the website statista.com, in 2015, the restaurant industry in the U.S generated $745 billion from sales of food and drinks. On average, U.S. residents bought 194 meals per person in 2011, which is less than 209 meals per person in the year 2000.

According to the website statista.com, the U.S. Gas and Oil industry revenue in the year 2014 was $220 billion.

According to the website statista.com, the revenues generated in private construction for the year 2014 in the U.S. was $718 billion.

According to the website statista.com, the U.S. private security industry generated $25 billion in 2014.

According to an article on the Washington Post's website, U.S. military spending was estimated to be around $597 billion for the year 2016.

According to statista.com, donations to various charity institutions in 2016 amounted to $390 billion. According to Forbes magazine, the ten biggest charities had revenue of over $20 billion in 2016, with United Way, Feeding America and Salvation Army having revenues of about $3 billion each.

According to techreport.ngo, there are over 10 million charity related organizations worldwide, whose combined revenues would make them the fifth largest economy in the world. The website also reports that there are 1.4 million charity related organizations in the U.S. that employ 11 million Americans. In India, there are 3.3 million charity related organizations, approximately one charity organization for every 400 people.

The institutions are today the main suppliers of many basic needs (food, energy, construction, security, etc.), but increasingly need less labor/skills (leaving people with nothing to trade with for the goods and services from these institutions), which raises the question, can this culture of "Institution Dependency" survive for much longer?

REDUCE YOUR DEPENDENCY ON INSTITUTIONS

During the "industrial revolution, a relationship between the institutions and the people was established. It involved humans trading their labor/skills with the institutions for money in order to enable them to buy the products and services sold by these institutions. With the disruption of technologies today, the institutions need increasingly less labor to produce their products and services, and as a result many people no longer have the means to make purchases from these institutions. At the same time, society had increasingly emphasized acquiring skills that enabled you to provide labor to the institutions to the detriment of skills that earlier populations had acquired in order to provide for themselves before the institutions existed (i.e. hunting, fishing, cooking, building shelter, security, etc.). The problem of "institution dependency" is more acute in developed/industrialized countries than in developing countries, which have only more recently industrialized.

In this publication, I want to discuss how an individual can get richer financially by reducing their dependence on institutions for their basic and other needs. This is especially important today because, the institutions are experiencing revenue problems as the purchasing power of the general population decreases due to the decreasing labor needed by the institutions. The institutions, which as demonstrate in a previous short publication, are making billions of US Dollars from the people's dependency on them for their basic needs and more, are fighting hard to keep the revenues high. They are using various tactics that in many cases are detrimental to the financial health of individuals whose purchasing power is on the decrease.

Here are some ways to save money currently going to institutions for basic and other needs:

How do you spend your luxury time and how much does it cost? Choose cheaper alternatives like spending time in a free park instead of an amusement/entertainment park that charges for entrance and rides. According to ibisworld.com, the amusement parks industry in the U.S. generated $16 billion in revenues for the year 2015.

Cook your meals "from scratch" at home (i.e. buy the food in its raw and most unprocessed state) to reduce the costs. According to the website statista.com, in 2014, the U.S. frozen ready dinner's market generated $15 billion in revenues. According to the website statista.com, in 2015, the restaurant industry in the U.S generated $745 billion from sales of food and drinks. On average, U.S. residents bought 194 meals per person in 2011, which is less than 209 meals per person in the year 2000.

Learn how to repair and renovate your house and its equipment as much as possible to save on having to pay somebody else (usually an institution) to fix it for you. According to the website statista.com, the revenues generated in private construction for the year 2014 in the U.S. was $718 billion.

Closely analyze the products and services you buy from the institutions. Today, many are designed to keep you paying more money as you use them or buy new ones after short periods of time. Examples include, "smart phones" that require new software and models every two years, video and music streaming services that cause you to pay per use, and also buy more internet data services, motor vehicles that increasingly need institution specialists to repair the most minor problems including a bulb change, credit card and bank accounts which use numerous devious tactics to get your money and social media tactics (many times using famous people) to promote products and services that have limited useful utility but appeal to your vanity.

Think of starting your own small business for extra income (if you already work at an institution) and/or working with multiple institutions (to avoid being totally dependent on one).

Plan for your old age years, independent of institutions. Millions today rely or plan to rely on pension funds controlled by institutions. The agents managing these institutions have their interest and those of the institution as their top priority, not you! There have been several cases where governments have used money from pension funds for other projects, with the assumption that contributions from younger generations will cover the short-fall, an assumption that was wrong given the decreasing employees in today's institutions, and thus decreasing contributions.

Institutions today increasingly need less labor and as a result have fewer opportunities available for individuals. This has led to fierce competition for these few opportunities from a growing population that still places a premium value on them culturally. The end result is that even those who access these opportunities will see their remuneration continue to fall as labor supply greatly outstrips demand. In today's information age environment, a specialized skill does not remain exclusive for long. The same goes with new innovations. The days when you "milked" an innovation or skill for lots of money for extended periods of time are long gone. The institutions have implemented their strategy for survival in the new economic environment using the tactics I describe above among others. It remains up to you to come up with your own strategy in this new economy and I provide some information below that can help you start formulating your strategies.

One of the fastest growing parts of the economy today are based on transactions that simply transfer money from one person to another, instead of the creation of new value via innovation (i.e. new products and services). Some people have labeled this fast growing segment of the current economy, the "Casino economy". I discuss this segment of the economy in the third short publication of the Kikulacho Cultures Series™ "The culture of "Easy Money" and your finances". This is an economy in which the institutions usually get the money, and individuals loose it. Beware!

In his 2010 book, "Empire of illusion", Chris Hedges states the following about the U.S., which equally applies to the rest of the world; "We now live in two Americas. One—now the minority—functions in a print-based, literate world that can cope with complexity and can separate illusion from truth. The other—the majority—is retreating from a reality-based world into one of false certainty and magic. To this majority—which crosses social class lines, though the poor are overwhelmingly affected—presidential debate and political rhetoric is pitched at a sixth-grade reading level. In this "other America," serious film and theater, as well as newspapers and books, are being pushed to the margins of society." I recommend this book for reading!

Chris Hedges is pointing out that, most people in today's world are too lazy and dis-interested in understanding the new complex world in which institutions will no longer plan and manage your life for you. After three centuries of people relying on these institutions to plan and manage their life (including expecting government to protect them from scams, instead of learning about them and avoiding them), they are too afraid to be independent and invest the time to acquire the skills required to live independent of the institutions. They find this too stress-full and instead try and escape from this reality. Preparing your strategies based on the new realities, can make you richer financially and enhance your chances of survival in this new world where you will be solely in-charge of your own destiny. Independence comes with both risks and rewards, and in today's environment is becoming increasingly the only choice.

COMPETE WITH THE INSTITUTIONS INSTEAD OF DEPENDING ON THEM!

In previous publications, I discussed how dependent a large portion of the population has become on institutions (Government/Public institutions, Private/Business institutions, and non-profit institutions) for their basic needs of food, shelter, energy, security, etc. I noted that this was a recent development which begun almost three centuries ago as a result of the "industrial revolution". This established a relationship between the institutions and the people, whereby humans traded their labor/skills with the institutions in order to enable them to buy the products and services sold by these institutions. With the disruption of technologies today, the institutions need increasingly less labor to produce their products and services, and as a result many people no longer have the means to make purchases from these institutions.

So what's next? Here is the good news, the same technologies that disrupted the "cozy" trading relationship between humans and institutions, has made it easier for the humans to compete quite effectively with the institutions. Technology has made it easier and cheaper for individuals to offer many products and services formerly only offered by the institutions. Here are some examples;

An article on the forbes.com website, cites an estimation by the market research firm Global Industry Analyst that in 2015 the online learning industry had $107 billion in revenues. Many online learning websites, which are not affiliated with institutions, are offering individual professional educators an opportunity to set up courses outside the institutions, including Lynda.com, Coursera.org and udemy.com. With the cost of education from institutions increasing, while the earning power of the graduates decreases, many students are looking for cheaper sources of education.

An article on the time.com website discusses the rapid growth of online doctor visits. 15 minute online doctor consultations cost between $40 and $50. Major Telehealth providers include, American Well, Doctor on Demand, MD Live and Teladoc. The article goes on to say; "Telehealth providers are reporting tremendous growth. American Well, for instance, pegs its growth at 1,100% in 2014 over 2013 and is going strong in 2015 so far, according to Chief Executive

Roy Schoenberg. Patient visits at Teladoc grew to 299,000 in 2014, from 127,000 in 2013, according to company documents. The company's current reach is 100 million people."

Individuals can leverage the new technologies to compete with the institutions, producing similar high quality goods and services at lower costs. Many institutions have high overhead costs, which are reflected in the costs of their goods and services. Individuals and small companies can minimize their overhead costs due to the many options available in today's new economic environment. I call this new economic environment, the Devolved Economic Environment. Just like devolved government, the business sector is becoming increasingly devolved in nature, with smaller businesses and individuals now able to produce goods and services that were once the preserve of the big business institutions. Even the big institutions today prefer to partner with entrepreneurs and consultants instead for employing direct employees.

In a survey carried out in 2014 by the Harvard Business Review, cited in an article appearing on the hbr.org website, titled "Half of Employees don't feel respected by their bosses", 54% of employees felt they don't get respect from their leaders. The competition for shrinking employment opportunities within the institutions and the desperation with which many look for these "coveted" positions, has made it possible for the lack of respect and mistreatment many are experiencing within these institutions. The leaders are aware that many of their employees are extremely scared of losing their employment, especially the higher paying position, and as a result may mistreat and disrespect them.

The dis-respect of employees today by the institutions is openly blatant, with vague and ridiculous employment requirements appearing in the advertisements that have de-emphasized objective skills such as education and training, and replaced them with the so called "soft skills" which are subjective, e.g. attitude, works well with others, etc. I am yet to come across an objective way of measuring attitude, and as for "working well with others" my experience has been it depends on who the others are! It gets even more ridiculous with job descriptions which include nonsense such as "should be able to work well under pressure", who does! The disrespectful environment that currently exists for many employees within the institutions has had many negative effects on them including health related problems.

According to an article on msn.com titled "A whopping 80 percent of Americans are in debt" discusses how of the 80% of American's in debt, 44% hold mortgages debt. The article goes on to say, "In fact, millennial's are much more likely to have student loan debt (41% have it), car loans (41%) or credit card debt (39%) than they are to have a mortgage. Among all other generations, mortgages are the leading component of consumer debt." For most people working within the institutions today, their home is owned by an institution (usually a bank) via a mortgage, their car is owned by an institution via the loan or lease arrangement, the retirement plan they pay into is controlled by institutions, their health-care access is dependent on institutions, they have credit card debt owed to banks and they owe an institution (usually the government) for their education via a student loan. For many, institutions touch on every part of their survival, making it seem impossible to provide for yourself outside the same. And as a result there is fear, in many, when they see the trend by the institutions that increasingly requires less labor.

Knowing that there are options outside the realm of employment with the institutions will put you back in control of your future endeavors, and free you from the dependency on institutions.

THE CULTURE OF CREATING COMPLEXITY; KEEP IT SHORT AND SIMPLE (THE KISS RULE)

Per discussions in my previous short publications, the new economic environment has short new product/new skill/new innovation life cycles and fast changing consumer needs which results in "brutal" competition, smaller profit margins, and falling wages. In this new economic environment, the relevance or real value of the goods and services traded becomes a focal point of many. However, many institutions are trying their best to avoid this reality by using anti-competitive practices and attempting to suppress the dissemination of knowledge and information that would erode the monopolistic advantages they have relied on for so long to survive. You can find many industries whose existence is heavily dependent on complicating things that could be made simple, in which case the industry would either shrink drastically, have greatly reduced profits/wages or cease to exist. These industries continue to have large revenues and their services continue to be critical. I will look at four dominant economic sectors to analyze this phenomenon; Health, Education, Information Technology and Justice.

Health

The content in all fields of education continues to grow rapidly, and in specific areas, namely the so called STEM areas (Science, Technology, Engineering and Math's) more complex. The complexity of the so called STEM courses discourages many from pursuing them. They have thus created a shortage in these fields, which raises the wages and profits in these sectors. Is this intentional? There are many who believe that if an educator knows their subject matter well, they should be able to explain it in a short and simple manner to anybody with a high school education. So what is wrong with the STEM courses? The inability of the educators in the STEM courses to present their material in a short and simple way could only mean that they either do not understand the material well enough to explain it simply, or they are creating an intentional shortage of personnel to keep wages and profits up in their professions. This is a problem in an economy that is headed in the opposite direction for most of the population, and puts their services out of the affordable price range. In-fact, in the U.S. today, most of the revenue in the expensive health sector comes from government funded programs like Medicaid, Medicare and government subsidized health insurance programs. Were it not for the government programs, this sector would collapse due to a lack of patients that can afford their services. These STEM programs need not be complicated, and should enable the mass spreading of their knowledge to decrease prices and make the services affordable based on the incomes of the vast majority of their potential customers. If this does not happen, the culture of complexity which is currently enriching them, will lead to their downfall, when the government can no longer afford to subsidize their customers. The U.S. health care sector revenues are over 1 trillion U.S. dollars, according to CSImarket.com.

Education

The content of education has grown rapidly, however, its utility is now frequently questioned. Is it possible that the growth of content is simply a way to charge students more money in an education system that sells it services in packages (i.e. Associates Degree or Degree) that require a set number of courses and content? How useful is this information to the student after graduation, and could some of it be acquired later through self-study?

In the U.S. most students over the last couple of decades have funded their education using government loans. With the changing economic environment, the default rate on these student loans is up according to an article in the Washington Post dated 28th September 2017 titled 'The number of people defaulting on federal student loans is climbing". Many college graduates in the U.S. no longer believe they will be able to repay their student loans, given their earnings potential in the new economic environment.

Is it time for colleges to change the amount of content, length of the programs and keep it short and simple? Alternatives are already being provided, including free online universities, self-learning books and You-tube videos some of which are greatly shortening and simplifying topics that take a whole semester to teach in colleges.

The education sector is made up of public and private schools. In the U.S., According to usgovernmentspending.com, public (tax and lottery) funded schools in the U.S. make up the majority with total spending in 2014 equaling $915 billion. The largest public school spending occurred in California $112 billion (12%), Texas $77 billion (8%) and New York $74 billion (8%). Private schools in 2014 had revenues of $54 billion according to statista.com.

Information Technology

In the U.S. the Information technology (IT) sector had 346 billion Euros in revenues while the European Union had 314 billion Euros in revenues. Many employers say there is a shortage of IT professionals. Is the content of the training presented in a complicated difficult to learn methodology? If so why? The wages and profits in the IT sector are also known to be very high compared to other sectors. The communication sector which relies heavily on IT products and services is also one of the most expensive sectors in the U.S., and is valued at $ 1.4 Trillion by statista.com.

Justice

Have you ever tried reading a law passed by the U.S. congress? These laws are long and difficult to understand! Is this intentional? Law schools in the U.S. are some of the most expensive, and consequently acquiring the service of a lawyer for your defense is equally expensive. However, trying to represent yourself in a justice system full of complex rules, laws and procedures, is very difficult if not impossible!

Essentially, the culture of creating complexity has captured another industry causing inflated costs, for a population with declining wages and profits! The largest revenues are generated in the corporate legal departments ($160 billion according to statista.com). According to Statista.com, the Legal Service industry in the U.S. is expected to generate a total of $ 288 billion in 2018.

The culture of creating complexity is costing everybody, by inflating prices, in an economic environment of falling wages and profits. Those in these sectors that are currently benefiting from higher wages and profits may not view this as a problem, however, when the institutions (government, corporate, etc.), which derive their revenues from the general population can no longer afford the inflated prices, the financial collapse in these industries will be massive. Another bubble waiting to burst!

THE CULTURE OF FEARING INDIVIDUAL EMPOWERMENT; VULNERABILITY PREFERRED.

After the recent mass shooting at a Florida school, the debate on the unique right that U.S. citizens have to bear arms is once again in the public domain. It is important to note that in most countries around the world, this right does not exist, ensuring that the governments in these countries have the sole right to purchase and bear arms. This right in the U.S. is enshrined in the Second Amendment of the constitution and according to Wikipedia; "The Second Amendment was based partially on the right to keep and bear arms in English common law and was influenced by the English Bill of Rights of 1689. Sir William Blackstone described this right as an auxiliary right, supporting the natural rights of self-defense and resistance to oppression, and the civic duty to act in concert in defense of the state."

The empowerment of the individual to provide independently for themselves, the necessities of life; food, clothing, shelter, security, health, education, etc.

has always been at the center of the struggle between individuals and institutions/organizations. The more an individual can provide these things for themselves, the less they need the institutions/organizations to provide it for them. In the past, institutions/organization have been used by individuals or a group of individuals to prevent the empowerment of the individual to provide various necessities of life; in order to maintain their dependency, oppress them and/or exploit them. The institutions/organizations have used a variety of strategies to instill a lack of self-confidence and fear in the general population in regards to their ability to empower themselves in ways that reduce their dependencies. In addition, a strategy by institutions to keep hidden knowledge and information that would empower the population, by using monopolistic tactics and their control of the various means of communication, has greatly slowed down the progress of individual empowerment (see my previous short publications on this website for the various specific tactics used and the dollar amounts involved). It is from this perspective that I evaluate the ongoing debate on gun control in the U.S.

The recent election of President Trump due to his popular policies against the "globalization" agenda and the vote by British citizens to exit the European Union, another blow to the "globalization" agenda, are a good place to start this evaluation. I monitored the media reporting on these two historic votes and noted a similarity; specifically, most media outlet personalities reporting the news had a hostile bias to the outcomes. It was clear that their independence was compromised by either personal or institutional interests. It is no secret that the "globalization" agenda is pushed by institutions/organizations and wealthy individuals. Voters who oppose "globalization" are therefore opposing the interests of the individuals that benefit from these institutions/organizations.

In many countries around the world, the problem of voters who vote against the interest of the institutions/organizations and wealthy individuals has been resolved by imposing "puppet" dictatorships that support their "globalist" masters. This is increasingly done via the rigging of elections against the majority's wishes, and when protests occur, they are silenced using armed forces from the government institutions. Since arms are the monopoly of the government, there is very little resistance encountered. This would be difficult in the U.S. given the second amendment which not only guarantees an individual's right to bear arms, but allows the decentralization of the armed forces, with states and local authorities having their own armed militias in the form of the National Guard, County Sheriff departments and Municipality Police departments. Many people in the U.S. believe that the gun control debate is about dismantling this unique right, because it has made it difficult for the "globalists" to neutralize the growing dissatisfaction with "globalization among the U.S. citizens. Using of the tactics utilized in other countries (e.g. rigging elections), would be a challenge with an armed population.

Empowering the individual to provide for themselves outside of the institutions, and thus end their dependency, is a central theme of my companies website. I focus on providing information on the tools and methods that will allow you to achieve this goal. Specifically, the "Kikulacho Cultures Series™" discusses lifestyle choices/cultures that are intentionally promoted by the institutions that dis-empower individuals to the advantage of the institutions. In a post on the business networking site, LinkedIn, Prof. Ngarua advised the following; "Empowering yourself to successfully resist those that empower themselves via dis-empowering others is the key to a wonderful life experience".

Apart from "the for profit institutions" that benefit from individuals being dis-empowered and depending on their products and services, non-profit organizations are also beneficiaries.

According to statista.com, donations to various charity institutions in 2016 amounted to $390 billion. According to Forbes magazine, the ten biggest charities had revenue of over $20 billion in 2016, with United Way, Feeding America and Salvation Army having revenues of about $3 billion each.

According to techreport.ngo, there are over 10 million charity related organizations worldwide, whose combined revenues would make them the fifth largest economy in the world. The website also reports that there are 1.4 million charity related organizations in the U.S. that employ 11 million Americans. In India, there are 3.3 million charity related organizations, approximately one charity organization for every 400 people.

Over the last three decades, a new "economic class" has emerged that benefits from others being dis-empowered. They support the efforts of the institutions to use various methods to thwart any attempts by the population to reduce their dependency on the institutions which they benefit from financially either via employment or ownership. Members of this class are expected to adopt and promote narratives, ideas and cultures that support the "globalization" agenda and continued dependence on institutions by the population.

The right to bear arms in the U.S. ensures that an individual can provide security for themselves outside of the institutions, and is therefore in line with the empowerment of the individual.

THE CULTURE OF "PACKAGED THOUGHTS"; WHICH HAS PRODUCED LAZY THINKERS

In this publication I discuss the popular culture of "packaged thinking". Today we live in a world where many people readily absorb packaged thoughts, designed and distributed by institutions. These packaged thoughts determine their opinions on specific issues. People self-identify with packaged thoughts such as, liberalism, conservatism, socialism, capitalism, feminism etc., which tell them what their opinions should be on various specific issues, without them researching and analyzing the specific issues separately. Any attempts to deviate from the packaged opinions within these self-identifying groups, draws the wrath of the rest of the group, with the social pressure threat of no longer being identified with the group being swiftly applied. This has led to a society with diminished critical thinking skills, since once a person identifies with a group, they simply follow the prescribed opinions blindly without much research and analysis.

This has enabled the growth of established political parties, religious institutions and corporate cultures that tell people how to think, act, speak, look and relate to others. These packaged thoughts eventually influence the purchasing, voting and life style choices of many people, which is the aim of the institutions that push the packaged thoughts. The main leverage these institutions have had for pushing their agenda is the fact that for a long time large portions of the population earned a living either directly or indirectly from them. With that changing, many people no longer feel the need to "toe the line", and an increasing number are finding independent research and analysis on certain issues is crucial for their survival in the new economic environment. As you can expect, most of the packaged thoughts were favorable to the institutions that promoted them, and did not take into account the needs of the individuals, especially when they found themselves no longer earning a living from the institutions.

This has led to an emerging growth of independent minded people, who are now more willing to analyze specific issues independently, without following the packaged prescriptions. The results can be seen currently, especially in the western countries. The clearest example is the U.S. electorate's election of a candidate who was opposed by the establishments of both major parties, the architects of packaged so called "liberal" and "conservative" thoughts. President Trump's campaign failed to follow the packaged thoughts of the Republican Party he was associating with, drawing the wrath of the establishment but winning the election. He also refused to think, act, speak, look and relate to others as prescribed by the political establishment, both during the campaign and since becoming President!

Here are some examples of the biggest financial beneficiaries of packaged thoughts, who will be the biggest losers as things change.

Political parties in many countries sell packaged thoughts to the electorate. According to electoralcommission.org.uk, in 2014, the Labor Party had an income of 39,570,000 British Pounds while the Conservative and Unionist Party had an income of 37,446,000 British Pounds. Both parties also spent almost all their income during that year. According to opensecrets.org, the U.S. Democratic Party raised $1,292,188,200 for the 2016 election, while the Republican Party raised $968,828,665 during the same year. During the 2012 election, the Democratic Party raised $1,069,752,788, while the Republicans raised $1,022,488,047.

According to an article appearing on the website theguardian.com titled "Religion in US worth more than Google and Apple combined" the business of selling packaged ideas via religious institutions commonly referred to as "religion", generates revenues of $ 1.2 trillion annually. The article states; "The faith economy has a higher value than the combined revenues of the top 10 technology companies in the US, including Apple, Amazon and Google, says the analysis from Georgetown University in Washington DC.". "Twenty of the top 50 charities in the US are faith-based, with a combined operating revenue of $45.3bn." The religious institutions also own schools and hospitals in many countries including the U.S., which generate revenues in the hundreds of million dollars. These institutions are tax exempt in the U.S. and many other countries.

Lobby groups that sell packaged thoughts/ideas to decision makers (e.g. members of congress) and the general public (voters) are big business. According to the website opensecrets.org, the lobby groups in the U.S. generated revenues of $ 3.15 billion in 2016. These lobby groups sell packaged thoughts/ideas with the goal of influencing the direction of policy and laws in countries around the world.

THE CULTURE OF DISTRACTIONS; AVOIDING YOUR REALITIES VIA GOSSIP, DRUGS, SPORTS, ENTERTAINMENT, CEREMONIES AND ALCOHOL

A growing number of people around the world are willingly seeking various distractions to avoid the realities of their lives. The evidence is in the amount of time they spend on the various distractions, which is an indication of an addiction. The most common distractions include; religion, rituals, ceremonies, sex, gossip, drugs, sports, entertainment and alcohol. Many people engage in these distraction activities even when at work, causing many employers to put in place procedures and rules that attempt to stop these behaviors. On the road, distracted driving is widespread, endangering the safety of others. And when walking, it is now common to see people bumping into others and even tripping and falling because they are not paying attention to what is in-front on them, but instead are staring at their smart phone screens! Addiction is widely considered to be an indication of a person's attempts to avoid dealing with a perceived problem.

Many businesses have capitalized on this phenomenon, providing the distractions sought for a fee. In his best-selling book, "Empire of Illusion", Chris Hedges discusses how sports and entertainment are used to provide an alternative reality for many, helping them conquer (in their imaginations) their perceived obstacles via the actors and sportsmen. In reality, however, their perceived obstacles and problems remain, causing them to increasingly spend more time in their alternative reality. This works out well for the businesses providing these distractions, as their revenues grow.

Here are some examples of businesses that are benefiting from the growing culture of distractions:

According to an article on Forbes.com titled, "Sports industry to reach $73 billion by 2019", the sports market which was worth $60.5 billion in 2014, will be worth $73.5 billion by 2019. This is mostly because of "media rights deals", which means the number of people watching sports via various media is growing rapidly. Licensing of sports clothes via royalties by the owners of the sports team was a $698 million from the sale of clothes worth $12.8 billion in 2014 according to another article on Forbes.com titled "Sports licensing soars to $698 million in royalty revenue".

According to statisticbrain.com, 11,000 adult films are released every year in the U.S. and the annual industry revenues now stand at $13.3 billion. 55% of guests in hotels in the U.S. will order adult films, while 87% of college students in the U.S. have had phone or webcam sex. Also, the number one internet search term is sex.

According to hootsuite.com, social media advertising budgets worldwide have doubled from $16 billion in 2014 to $31 billion in 2016. The author projects a 26.3% global increase in 2017. Social media content can be categorized as mostly entertainment, sports and gossip within personal networks.

Marketresearchstore.com in its 2016 publication titled, "Global depression drug market set for rapid growth, to reach around 16.80 USD billion by 2020" notes the following: "Zion Research has published a new report. According to the report, the global depression drug market was valued at USD 14.51 billion in 2014 and is expected to generate revenue of USD 16.8 billion by end of 2020, growing at a CAGR of 2.50% between 2015 and 2020".

According to an article on economist.com titled "Celebrities endorsement earnings on social media" according to captiv8, an analytics platform that connects brands to social media, "someone with 3m-7m followers can charge, on average, $187,500 for a post on YouTube, $93,750 for a post on Facebook and $75,000 for a post on Instagram or Snapchat".

There is also a direct relationship between the rise of ceremonies and rituals in a community and the progressive decline of the majority of the people's ability to provide for themselves. During the transition period, when the decline begins, ceremonies and rituals increase rapidly as the elite beneficiaries of the system try to distract the majority from analyzing the cause of their declining fortunes.

The culture of distraction is costing you money that you could invest for future returns. Visit habarishujaa.com to find out how. Challenge your assumptions, ASK WHY!

THE CULTURE OF "SOCIAL SIGNALING" AKA "IMAGE", AND YOUR FINANCES

Some time ago, there was a popular English language saying; "Do not judge a book by the cover"! In today's society, it is increasingly popular to do just that. In-fact there is now a popular saying "Image is everything"! Times have changed, and as a result the business of image making is booming, and society is paying for it. Here are some figures.

Public relation firms, which concentrate on improving the image of an individual and/or organization (regardless of the reality), are big business today. According to Ibisworld.com, public relations firms in the U.S. generated $14 Billion in revenues and are projected to grow by 2.9% annually for the next five years. The industry employs over 92,000 people and has over 44,000 businesses involved.

According to brandstrategyinsider.com, over $466 billion was invested globally in 2011 for branding of products and services. With many products and services selling for a premium price due to branding, a large portion of the world's value today is based on branding/image, with the underlying tangible assets making up a very small portion of the valuation.

Many companies also use the "halo effect" (the tendency for an impression created in one area to influence opinion in another area) in advertising and marketing. Two categories stand out, sexual desire and celebrity images. Companies such as Victoria Secrets which in 2015 had revenues of over $ 7 billion, Calvin Klein with 2010 revenues of $ 2.5 billion and Playboy ($ 215 million in 2010) use images associated with sexual desire to sell their products.

According to an article on economist.com titled "Celebrities endorsement earnings on social media" according to captiv8, an analytics platform that connects brands to social media, "someone with 3m-7m followers can charge, on average, $187,500 for a post on YouTube, $93,750 for a post on Facebook and $75,000 for a post on Instagram or Snapchat".

Many of the most valuable companies in the world have the dollar value of their organization largely based on their brand/image. Examples include Facebook and Amazon. Many designer clothing lines are also based on the same model, where consumers pay a premium for the brand name, which may not necessarily mean they are getting a better quality product.

Large numbers of people invest considerable time creating an image meant to send a non-verbal message to others, i.e. social signaling. This is used to gain favor in a world where the culture of subjective decision making is popular. Pictures posted on Facebook, Instagram and prominently displayed in their homes and offices are intended to signal belonging to a group and or identifying with ideas, e.g. wealthy class, religious organization, married etc.

Many people are spending money on products or services that charge a premium for a brand name with no additional value compared to products and services that cost less. This money could be better used in other investment ventures. The consumer's ability to distinguish between the "hype" of a brand name and real value, could make a difference in one's financial well-being.

THE CULTURE OF INFLATION; EXPENSIVE EXPECTATIONS SHRINKING INCOMES

In 1992, Janet Jackson released the "hit" song titled, "the best things in life are free". It was a soundtrack to the 1992 movie "Mo Money". Today's popular culture seems to disagree with the song title but agree with the movie title! The evidence of this is clearly displayed with the most popular kind of entertainment themes today. The most popular movies, songs and tabloid stories have the across the board theme of the best things in life being expensive and requiring lots of money (more in line with Colloway's 1989 "hit" song "I wanna be rich").

This culture of inflation, fueled by expensive expectations, is facing a "headwind" of decreased purchasing power around the world. According to World Bank statistics, the Labor Participation Rate around the world has been falling and is now at 62% compared to 67% in 1995. In the U.S., the overwhelming majority of the jobs created in the last several years have been close to minimum wage jobs, while the number of high paying jobs are shrinking fast. According to the U.S. census bureau, per capita income fell from $29,000 in 2006 to $27,000 in 2011.

In a normally functioning free market economy, this should lead to a depression. With purchasing power of the population rapidly declining, there should be a lack of demand for goods and services and the lowering of prices leading to a depression. This however has not been the case! Why?

The institutions, Government, big business and non-profit/religious institutions would be the hardest hit by a depression. The beneficiaries of a depression would be the population whose purchasing power is decreasing every single day as prices remain the same or actually go up! A 19th July 2016 article appearing on cnn.com titled "I pay more than half my income in rent", states the following; "Millions of Americans are getting squeezed as rents rise much faster than wages. A recent report from Harvard's Joint Center for Housing Studies showed that 11 million people are spending more than half their income on rent".

However, the popular culture of inflation prevents many from seeing the contradiction between falling wages and continued rising prices. Instead of asking why prices are not falling, they ask why their incomes are not increasing! As I discuss in my other short publications, the new economic environment is likely to see a continuous decrease, not increase in wages and profit margins. Increased wages and profit margins only happen in an environment of scarcity, an environment we currently don't have due to excess available labor, goods and services.

The institutions have however created their own artificial shortages via monopolistic and other anti-competitive practices, as well as promoting a culture of inflation. With institutions in control of the media, entertainment, education and communication channels, the culture of inflation is promoted by making expensive expectations popular. This is necessary for the survival of the institutions which have large overheads that cannot survive in an environment of low profit margins. The populations have responded positively to the culture of inflated expectations, which have seen them borrow money to close the gap between their falling incomes and their spending. According to nerdwallet.com, the average household credit card debt has increased by 11% in the last decade to about $16,000, and when you add the mortgage, it stands at $134,000. Car and student loan debt is on the rise and is predicted by many economists to be the next bubble to bust in the U.S.

The culture of inflation also affects one's ability to invest, something that is increasingly crucial in today's economic environment. With most of your resources going to inflated consumption, you cannot exploit the numerous investment opportunities available in today's increasingly devolved economy, which allows entrepreneurs to thrive. Visit habarishujaa.com to discover how to free yourself from the culture of inflation.

THE CULTURE OF "EASY MONEY" AND YOUR FINANCES

Today, the culture of "Easy Money" is very popular. Many people talk about longing for the day when they can "sit back and watch others make money for them". As a result, a significant portion of today's economy is made up of industries that attract those longing for "Easy Money", such as the Gambling industry (lotteries, casinos, sports betting, etc.) and the Pyramid Scheme industry (e.g. Multi-level marketing, etc.).

In the U.S., the largest lotteries are owned by the State Government's, which use it as a secondary source of revenue (their primary source of revenue being taxes). Individuals in the U.S. spend over $70 billion on lottery tickets, and many are purchased by people making incomes below the poverty rate. The chance of an individual winning the lottery is one in 175 million in multi-state power-ball lotteries according to wikihow.com. The lottery industry is growing rapidly around the world, and many of the participants are the poor.

The sports betting industry is also rapidly growing. According to the website, statista.com, the sports industry worldwide is estimated to generate revenues of over $ 700 billion, with the fastest growing segment being online betting.

The Pyramid Scheme industry, which today is mostly practiced via Multi-level marketing schemes, is estimated to have generated revenues over $31 billion in the U.S. according to the website mlmwatch.org. Companies such as Amway, are big businesses with tens of thousands participating in the U.S. While a few people make large amounts of money, most lose money, which is the nature of pyramid schemes (and thus the name).

THE CULTURE OF FEARING UNCERTAINTY, AND THE FALSE SOLUTIONS ADOPTED

"Uncertainty is the only certainty there is, and knowing how to live with insecurity is the only security." (John Allen Paulos)

"The outline of our lives, like the candle's flame, is continuously coaxed in new directions by a variety of random events that, along with our responses to them, determine our fate. As a result, life is both hard to predict and hard to interpret." (The drunkard's walk by Leonard Mlodinow)

"In this life, security is a fallacy...., Overcome the "Permanence mentality syndrome". Even business stability should be challenged...employment has the ability to destroy your self-confidence as a creative being and gradually develop a dependent mentality." (You don't need a job by Dr. Kinyanjui Nganga)

The industrial revolution provided humans with the ability to provide solutions to uncertainties and insecurities that had historically been the norm. The institutions it created produced new innovative goods and services with large profit margins due to artificial scarcity or monopolistic protections. Unlike the past when know-how of technological innovations such as farming tools (hoes, rakes, the plows pulled by Oxen, etc.) were widely shared enabling others to develop their own using available natural resources without having to purchase it from the innovator, the institutions succeeded in monopolizing technological innovations by restricting the know-how and thus forcing the population to buy the products and services from them.

These large profit margins provided the institutions with the ability to offer the population new solutions to uncertainty and insecurities including, permanent/long term employment opportunities, salaried jobs, sick pay, disability pay, retirement pay, and large tax revenues for the government allowing for "social safety nets" and permanent government employees including professional politicians, bureaucrats and security personnel. The aim was to ensure dependency on the institutions, for a population that had for many years been independently providing for themselves via natural resources. The dependency was needed, in order for the population to be willing to trade their labor to the institutions, something that the institutions could not do without in their production operations. The institutions also needed a market for the goods and services they were producing, and a population that independently produced for themselves was not a good market. That era has now come to an end for two reasons. The first is that technological advances have enabled the institutions to rely less on human labor for production. The second is that the same technology is now allowing for the easy access to technological know-how enabling the population to go back to the former scenario whereby they could produce the good and services for themselves without buying them from the institutions.

The new economic environment has short new product/new skill/new innovation life cycles and fast changing consumer needs which results in "brutal" competition, smaller profit margins, and falling wages. In this new environment, uncertainty and insecurity are back, and many are unable to cope with the new reality. However, for those willing to change their mind-set and end their dependency on institutions, there are numerous new opportunities.

Realizing there are many people who do not want to accept their new reality, institutions and individuals have devised false solutions that allow them to exploit their vulnerabilities. The false solutions offered have allowed a few people to continue to enjoy large profits and salaries at the expense of those that want to deny their reality, despite the changed economic environment.

The false solutions include employment and investment opportunities that promise unrealistic future benefits for themselves and/or their dependents. In the field of employment, many new employees are enticed with promises of future larger salaries and benefits that never materialize. Their egos are massaged with titles meant to create the impression of career advancement, and power structures are put in place to encourage a culture of bullying and mistreatment of subordinates that allows for a release of frustrations and anger on others by the so called "bosses".

In the investment field, many are encouraged to invest heavily in their children's education, with the false promise of eliminating uncertainty and insecurity for them despite the new realities. Many of these schools (private schools and "prestigious" colleges) still prepare their students for the economic environment that no longer exits, but the owners and executives continue to earn profit margins and salaries that are increasingly abnormal in the new environment. According to usgovernmentspending.com, public (tax and lottery) funded schools in the U.S. make up the majority with total spending in 2014 equaling $915 billion. Private schools in 2014 had revenues of $54 billion according to statista.com.

Pyramid schemes, multilevel marketing schemes and fraudulent investment schemes are on the rise targeting those that will not accept their new realities. Motivation speakers that offer "soundbite" solutions where participants chant tired slogans in an attempt to temporarily escape their realities are very popular. Other forms of escapism including religion, alcohol, porn, sex, drugs, entertainment, rituals, etc., have become a major consumer of many peoples resources and time, a clear sign of addiction.

The financial sector and its derivative commodities (currency/money, stocks, bonds, insurance certificates, etc.) which were created for the sole reason of allowing for the smooth trade between people trading their labor in exchange for their ability to buy goods and services from the institutions, now has a new role. One of their new roles is that of providing credit facilities to a population with diminished purchasing power, in a way that keeps them dependent on the goods and services provided by the institutions. The second role is that of providing insurance and various investment products which provide people with a false sense of security, while ensuring that there are no guarantees that the benefits will materialize (the detail is in the agreements fine print!). According to selectusa.gov, as of the end of 2016, the U.S. banking system had $16.8 trillion in assets, while in 2015, the insurance industry's net premiums written totaled approximately $1.2 trillion. U.S. private equity firms invested more than $644 billion in U.S.-based companies in 2016 mostly using retirement funds of people who have little say on how their money is invested.

The false solutions have led to those targeted experiencing unprecedented psychological and physical health problems. This could be a major contributor to the fast growing health sector in the U.S. and around the world. According to marketresearchstore.com, the global depression drug market estimated at $ 14.51 billion in 2014 is set to experience rapid growth in the coming years.

The biggest fear of the institutions, is the discovery, by today population, of their ability to provide for themselves outside the realm of the institutions, thus reducing and eventually eliminating their dependency on them. Ngarua Services via its website habarishujaa.com is providing information on how making strategic lifestyle choices in line with the new economic realities can help you exploit the numerous opportunities available today, that allow you to free yourself from dependency on the institutions and their inflated needs. The website also provides business ideas categorized by sectors, which provide a starting point for freeing yourself from false solutions and successfully planning and strategizing in the new uncertain economic environment. Facing your uncertainties and insecurities, planning and strategizing realistically, and implementing realistic plans provides "Real Motivation" to your life.

THE CULTURE OF VALUING SUBJECTIVITY

Currently, the news is dominated by claims of sexual harassment in various work places. To-date, however, there has been no debate in regards to the kind of environment that allows harassment to be so prevalent in the work place. There are many kinds of harassment taking place in today's workplace, and they can be traced to a work environment that values subjective criteria over objective criteria in the decision making processes. The Cambridge dictionary describes the word Subjective as; "influenced by or based on personal beliefs or feelings rather than based on facts" and the word Objective as; "based on real facts and not influenced by personal beliefs or feelings".

Many of the decision making criteria today in the work place and other arenas are subjective rather than objective. They therefore allow the decision maker to assert their biases into the decision making process, and since "feelings" are more important than "facts", they have wide latitude in justifying their biased decisions. This in turn means that those seeking a favorable decision must conform to the expectations of the decision maker, even when the result is harassment.

We have not heard much in the news about those who resisted the sexual harassment advances. However, famous actress Lupita Nyongo did talk about her experience and fears, when she resisted the sexual advances of a powerful person in the film industry. It is quite clear from her story that she was aware her resistance could damage or destroy her career, an indication that it was a widely accepted practice for advancing ones career. In an environment where objective criteria are used by decision makers in choosing talent, this should not be the case! A subjective decision making process will always result in providing decision makers a wide choice of criteria with which to make their decisions allowing them to manipulate the process to their liking and needs.

Many people today want to be "favored" and support environments where favorable biases (i.e. subjectivity) result in gains for them. The excuses used to justify why they should be favored are provided by a belief they are superior and more deserving than others based on things such as religious, ethnic or cultural affiliations. Many will network with relatives and friends to canvas for jobs and opportunities, which can only happen in a subjective environment where bias can be inserted in the decision making process. It is quite common for people today to use ethnic, race and religious affiliations to gain favor in a biased environment. Shared ideology/ideas can also be used to bias the decision making process by many. It is now common to hear hiring managers and employees emphasizing recruiting for a "good fit" rather than skills. Of-course the "good fit" is in the eye of the beholder! This explains the resistance by many to an objective decision making process. However, they will complain when their expectations are not met in the subjective environment!

President Trump made an observation during a press conference recently discussing the alleged sexual assault by Judge Kavanaugh's, after his appointment to the U.S. Supreme Court. He noted that the country was entering a dangerous phase, where the principle of people being presumed innocent until proven guilty beyond a reasonable doubt was no longer respected. At the time the President was making this observation, Judge Kavanaugh's confirmation in the U.S. Senate was experiencing continued delays due to allegations of sexual assault which had never been reported to local law enforcement for investigation and potential prosecution, but were now suddenly emerging as a result of his nomination to the U.S. Supreme court. Supporters of the claimant have been using empathy for victims of sexual assault, to argue that it is not always possible for an alleged victim to follow the due process of the law in seeking justice. The due process of the law allows for both accuser and accused to be given objective due process, unlike a "he said she said" subjective process that allows the injection of victim sympathy. Would a victim of robbery with violence be allowed to use a similar argument after failing to report a crime to the appropriate authorities?

Bill Cosby, the famous actor and comedian, recently got convicted of a sexual assault incident that took place many years ago. The victim used the due process provided via the justice system, which provides that one is innocent until proven guilty. Another principal in the justice system envisions justice as being blind (as exhibited in many statues outside courts around the world), eliminating subjectivity including empathy.

It is important to note that unlike sexual harassment, sexual assault is a criminal offence, and is expected to be reported to the nearest local enforcement as soon as it happens to allow for investigations in a timely manner. Failing to do this, jeopardizes the collection of evidence in a timely manner which is crucial for any credible and objective investigation.

Many of the actions taking by authorities once sexual assault is reported are very supportive of the alleged victim, including ensuring the accused is prohibited legally from coming into contact with the victim especially during the investigations stages.

According to an article in the press, while leaving the hearing of Judge Kavanaugh's sexual assault hearings, a U.S. Senator was confronted by a woman claiming she had been sexually assaulted and could not remember the details, and asking for his reaction. The Senator responded that the alleged victim should report the crime to the local police. The article condemns the Senator for this reply! The aim seems to be to promote an alternative subjective justice system based on the loudest mob, in what is commonly referred to as "mob justice".

This is not the only legal issue where "victim empathy" is being used as an excuse for not following the due process of law. The illegal immigration problem in the U.S. is based on this very excuse.

Advocates of illegal immigration claim that economic hardships and reports of criminal activity in the countries the illegal immigrants come from, should make them exempt from following U.S. immigration laws! The illegal immigrants also increasingly endanger children by involving them in their endeavors to enter to U.S. illegally, but the "globalists" believe that these children are another reason for the immigrants to be exempt from following the U.S. laws. This has allowed the population of illegal immigrants in the U.S. to grow rapidly since the 1990's to an estimated 11-20 million individuals. The supporters of illegal immigration, which is part of the "globalist" agenda, use the culture of "victim empathy" as well as claims of racism, to advance their open borders agenda. Illegal immigration has been found to cost the U.S. tax payer hundreds of millions of dollars, while reducing wages and or displacing legal workers.

The cost for the many institutions and companies that are allowing a subjective environment to prevail is now beginning to emerge. Millions of dollars have been spent to silence the victims of sexual harassment, and millions more are being spent on public relation campaigns to mitigate the damage on the image of these companies. The Weinstein Company, which was founded by Harvey Weinstein who has been widely accused of sexual harassment, is in talks to be sold in a bid to avoid its demise. The U.S. media company CBS recently removed its CEO Les Moonves amid allegations of sexual harassment, which are likely to cost the company millions of dollars in litigations and settlements. Other costs to institutions include incompetent and/or demotivated employees, which affects productivity and competitiveness.

The cost on the individuals suffering a negative impact from this subjective environment ranges from health and mental issues, which require medical treatment, to loss of income in situations where the biases do not favor them.

THE CULTURE OF ENFORCING "MORALITY"

In this publication, I discuss how society enforces so called "morals" and the financial impact on individuals and society. According to oxforddictionaries.com "morality" is defined as "Principles concerning the distinction between right and wrong or good and bad behavior". According to Paul Ree in his publication "Origin of Moral Feelings", different societies might have different moral values; in every society, right and wrong were defined by the society's needs and cultural conditions. In Friedrich Nietzsche's "The Genealogy of Morals", a response to Ree's aforementioned publication, Nietzsche shows how relative "morality" is, by giving examples from historical events in Europe where the "powerful/elite" used "morality" as one of their tools to exercise power over the rest of the population. In the book, "Might is Right" by Ragner Redbeard, a similar argument is presented.

Despite the relativity of "morality" many societies continue to enforce it via laws, economic discrimination and social pressure. A commonly used name for "morality" enforcement laws is "victim-less laws". Historically, laws were established within communities to avoid victims personally punishing those that harmed them, what is referred to as "taking the law into your own hands" or "the law of the jungle". By having common laws and institutions that enforce them, those harmed could seek justice via this system which is supposed to be a fair and unbiased arbitrator. The use of this system to enforce "morality" via victim-less laws can negatively harm the reputation and acceptability of this system as we will see below.

Here are some examples of "morality" enforcement laws, and how they differ between societies. According to Wikipedia, the age of sexual consent in various countries varies from 14 years (Italy, Germany, Portugal, etc.) to 18 years (Turkey, Malta, etc.). In the U.S. the age of sexual consent ranges from 16 years

(New Jersey, Pennsylvania, etc.) to 18 years (Florida, Virginia, etc.). "Statutory rape" rape charges are invoked in accordance with the established age in the relevant society. Wikipedia lists Brazil, Ethiopia, India, and about 30 other countries where prostitution is not illegal, but the vast majority of countries in the world have laws that prosecute those participating in prostitution activities. In many countries public nudity is forbidden outright on the basis that nudity is inherently sexual. Many states of the U.S. fine offenders on that basis. However, in many parts of Africa and Asia, traditional clothing may leave exposed many parts of the body that could be considered public nudity or indecent exposure in other countries, e.g. women's breasts. In today's internet and social media environment, sending nude pictures over the internet, commonly referred to as "sexting" has led to teens exchanging nude photos in the U.S. being charged under anti-pornography laws that prohibit naked images of those under the age of 18 years. On the issue of illegal drugs, over 26 states in the U.S. have now legalized Marijuana to be consumed in various formats, but many other state and countries still have laws that punish those who use Marijuana. Whereas in the U.S. it is legal to film, sell and watch pornographic movies, in many countries in Africa and Asia this is illegal by law. At the same time polygamy is legal in many African and Middle-East countries, but illegal in the U.S. and Europe. Sexual relations between people of the same sex is legal in the U.S. and many European countries but illegal and punishable by law in many African and Middle-East countries. The financial cost of these "morality" law enforcement efforts can be enormous as we will see in some examples below. The psychological confidence of the common law system by those punished by the system for "morality" inspired laws is greatly diminished.

Furthermore, the individuals and groups (e.g. religious institutions, elitist cliques, etc.) that enforce "morality" via economic discrimination (where those that do not conform are denied access to economic opportunities such as jobs and business contracts, usually through the now common practices of networking/canvasing for jobs, referrals and "crony capitalism") and community/relatives social pressure, create economic environments that are anti-competitive and lacking in diverse ideas, which kills innovation, the main ingredient in the expansion of economic opportunities. According to a 2009 survey by Gallup published on their website gallup.com, titled "Religiosity highest in world's poorest nations", the survey found that the more religious a country was, the poorer it is economically. Religion is a main transmitter and enforcer of "morality" in many societies today.

Below I look at some costs of enforcing victim-less laws in the U.S.;

According to an article on moneycation.com, a report analyzing the cost of prostitution control was conducted in the 1980s and revealed that the average arrest, court and incarceration fees, amount to approximately $2,000 per person arrested. The article also says that in 2010, 62,668 people were arrested under the laws criminalizing prostitution, which when adjusted for inflation would be a cost to society of approximately $ 206 million. Due to the enforcement of this "morality" law in the U.S., the cost for an individual seeking prostitution services in the U.S., the article points out, is $25 for a Chicago sex worker, while in South Africa which has less enforcement emphasis it is $1 an hour.

According to an article on the website suntimes.com, in the year 2015, the U.S. federal and state governments had spent more than $36 billion in the war on illegal drugs. According to the U.S. Federal Bureau of Prisons website, bop.gov, 46.3% of the inmate population is made up of individuals convicted of illegal drug related crimes (i.e. over 81,000 individuals). Essentially, almost half of the U.S. federal prison population is due to illegal drug related crimes!

According to a 2010 article on npr.org titled "States struggle to control sex offender costs", "Sex criminals, along with drug offenders, are the fastest-growing part of prison populations" in the U.S. The list of "sex criminals" includes; public urination (indecent exposure), "sexting" (minors sending nude pictures to each other via cell phones) and "Romeo and Juliet" cases involving older teens who had consensual sex with younger ones. The cost of incarceration and the tracking of these "offenders" via a legally mandated sexual offenders database is in the hundreds of millions U.S. dollars.

As pointed out above, individuals punished for "morality" inspired laws may decide the justice system is unfair and nonsensical, which could lead to them lashing out at society by committing "non-morality" inspired crimes which do have victims, since the consequences are similar or the same.

The money spent enforcing these "morality" laws, could instead be used to help prevent crimes with victims and/or provide a better justice system for crime victims.

IS MARRIAGE GOOD FOR YOUR FINANCES?

We live in a society where an arrangement known as marriage is encouraged. The arrangement is considered an important part of the culture in many societies today. Even though the way it is practiced differs slightly from one society to another, there has been an increase in conflict between the individuals involved in this arrangement known as marriage. The religious institutions are the main promoters of marriage and benefit financially from the practice. I believe it is time to start evaluating this arrangement in terms of its financial viability on both individuals and society. Here are a few examples of the negative financial impact of this popular culture:

According to the U.S. Center for Disease Control, Costs of intimate partner violence (IPV) against women alone in 1995 exceeded an estimated $5.8 billion. These costs included nearly $4.1 billion in the direct costs of medical and mental health care and nearly $1.8 billion in the indirect costs of lost productivity. This is generally considered an underestimate because the costs associated with the criminal justice system were not included.

In 1995, victims of severe IPV lost nearly 8 million days of paid work-the equivalent of more than 32,000 full-time jobs-and almost 5.6 million days of household productivity each year.

An article in 2009 by David G. Schramm, Ph.D. provided an overview of the estimated economic costs of divorce to individuals, communities, and state and federal governments, which he pegged at $33.3 billion annually in the U.S.

The prospects of a combined income that comes from this arrangement known as marriage may be attractive, however, when things go wrong the losses may exceed any gains made. As you plan your financial future, take into consideration the increasingly destructive nature of the marriage arrangement on people's finances. Many financially stable individuals have ended up in poverty because of this arrangement, including many well-known millionaire celebrities in the U.S.

Who benefits financially from marriage?

In a free market economy, individuals make choices and markets respond to the demand. Many of the fastest growing industries today are driven by popular cultures, so this is a big part of the world economy today. It is up-to the individual to be informed on both the short and long term satisfaction they will derive from the life choices they make.

The arrangement known as marriage (also commonly referred to as the

"Institutions of marriage") is currently a popular cultural practice in all communities in the world. The first short publication of the Kikulacho Cultures Series™ titled "Is marriage good for your finances" discussed the potential negative impact of marriage on an individual's finances. This publication will discuss who benefits financially from marriage, i.e. where your money goes.

According to a April 2016 IBISWorld.com report, the Dating Service Industry (e.g. dating sites) in the U.S. generates annual revenues of $2 billion with an annual growth rate of 4.5%., with over 4000 businesses that employ over 8000 people.

A June 2016 IBISWorld.com report states that the wedding service industry (i.e. wedding planning, catering, photography, limo-service, etc.) in the U.S. generates annual revenues of $72 billion with an annual growth rate of 3.2%, with over 300,000 businesses that employ over 1 million people.

A February 2016 IBISWorld.com report looks at the Marriage Counseling industry (i.e. Psychologists, Social Workers, Marriage Counselors, etc.) in the U.S. which generates annual revenues of $15 billion with an annual growth rate of 3.4% with over 140,000 businesses employing over 230,000 people. It is important to note that this excludes marriage counseling services performed by non-profits such as religious organizations which derive their revenues indirectly via donations. In most societies, religions are the major promoters of the marriage arrangement!

The Divorcestatistics.info website reports on the cost of divorce in the U.S. stating that "Divorce is a big business in the United States. It is in fact a $28 billion-a-year industry with an average cost of about $20,000."

So, there you have it, these are the financial winners of the arrangement known as marriage. It is big business, but some people's quality of life and finances may be ruined by a choice that is promoted heavily as attractive but is turning out to be disastrous.

Chapter 11 - Conclusions

The "globalists" elite agenda is no different from other elitist groups of people (e.g. royal families, bourgeoisie, religion based governments, etc.) that have tried over many years to promote a class based pyramid structured economic system that favored them at the top.

This agenda was severely interrupted in Europe by the "Enlightenment" period which changed the culture of a dependent population, to a culture that promoted individual freedoms and choices and emphasized independent thinking. The "globalists" are intent on reversing this culture, as it does not bode well for their continued financial privilege.

In his address to the United Nations General Assembly on the 25th of September 2018, President Trump said the following in regards to the "globalist agenda" and immigration;

"We will never surrender America's sovereignty to an unelected, unaccountable global bureaucracy. America is governed by Americans, we reject the ideology of globalism and we embrace the doctrine of patriotism. Around the world, responsible nations must defend against the threats to sovereignty not just from global governments but also from other new forms of coercion & domination."

"Illegal immigration exploits vulnerable populations, hurts hardworking citizens and has produced a vicious cycle of crime violence and poverty. We recognize the right of every nation in this room to set its own immigration policy in accordance to its national interest. Just as we ask other countries to respect our own right to do the same, which we are doing."

"That is one reason the United States will not participate in the new global compact on migration. Migration should not be governed by an international body unaccountable to our own citizens. Ultimately, the only long term solution to the migration crisis is to help people build more hopeful futures in their home countries. Make their countries great again."

Once again, I would like to reiterate that ultimately, the responsibility lies with the population to see the "globalist" agenda in light of the impact it will have on their culture and life-style choices and make decisions appropriately. Individuals need to value being in control of their destiny, for this is what is at stake. The "globalists" want to create a new kind of slavery, where the population's choices are made by an elite "globalist" group. This book analysis the "globalists" agenda and gives the population a chance to choose wisely.